KILLERS
INSECTS

PHILIP STEELE

Julian Messner

Copyright © 1991 by Julian Messner

First published by Heinemann Children's Reference,
a division of Heinemann Educational Books Ltd
Original Copyright © 1991 Heinemann Educational Books Ltd

All rights reserved including the right of
reproduction in whole or in part in any form.
Published by Julian Messner, a division of
Silver Burdett Press, Inc., Simon & Schuster, Inc.
Prentice Hall Bldg., Englewood Cliffs, NJ 07632

JULIAN MESSNER and colophon are trademarks of
Simon & Schuster, Inc.

U.S. project editor: Nancy Furstinger

Printed in Hong Kong

Lib. ed. 10 9 8 7 6 5 4 3 2 1
Paper ed. 10 9 8 7 6 5 4 3 2 1

Library of Congress Cataloging-in-Publication Data
Steele, Philip.
  Killers: spiders and insects/by Philip Steele.
    p. cm.
  Includes index
  Summary: Describes dangerous insects and spiders that sting or
bite, spread disease, or destroy crops.
  1. Insects – Juvenile literature. 2. Spiders – Juvenile literature.
  [1. Insects. 2. Spiders.] I. Title.
QL467.2.S775 1991
595.7 — dc20

ISBN 0-671-72235-2 ISBN 0-671-72236-0 (pbk.)   90-24519
                                                      CIP
                                                      AC

Photographic acknowledgments
The author and publishers wish to acknowledge, with thanks, the following photographic sources:
a above *b* below *l* left *r* right
Cover photograph courtesy of Oxford Scientific Films
Bruce Coleman pp12*a* (P A Hinchcliffe), 15*a* (F Sauer), 24 (Jane Burton), 29*b* (Peter Ward); Mary Evans Picture Library p31; Eric and David Hosking pp9, 14*a*, 16*r*, 19, 23, 26, 29*a* Frank Lane Picture Agency pp6, 8, 13; Mansell Collection p25; NHPA pp11 (S Krasemann), 12*b* (Otto Rogge), 17 and 30*a* (Anthony Bannister), 15*b*, 20, 21, 23, 27 (all Stephen Dalton); Shell Photograph Library p14*b*; Survival Anglia p16*l* (David Shale); Daily Telegraph Colour Library p30*b*; C James Webb p22*a*.
The publishers have made every effort to trace the copyright holders, but if they have inadvertently overlooked any, they will be pleased to make the necessary arrangement at the first opportunity.

CONTENTS

CREEPERS AND CRAWLERS	6
Biters, chewers, and suckers	7
Small but deadly	7
YOUNG AND DANGEROUS	8
Spoilers of food	9
Poison caterpillars	9
DEADLY STINGERS	10
Why do wasps and bees sting?	10
Warning colors	11
Killer bees	11
SOLDIERS ON THE MARCH	12
Army life	12
Columns of death	13
The demolition squad	13
TIRELESS JAWS	14
The potato-eaters	14
Poison beetles	15
Kitchen pests	15
DESTROYERS	16
Crop strippers	16
Life in the swarm	17
Pest control	17
DIRT BREEDS DANGER	18
Maggot-breeders	19
BLOODSUCKERS	20
The scourge of Africa	21
BUGS AND LICE	22
Strange bedfellows	23
THE PLAGUE FLEAS	24
A deadly disease	25
POISON SPIDERS	26
Hairy heavyweights	26
The deadliest of all	27
A STING IN THE TAIL	28
The poisoners	29
INSECT FRIENDS	30
Let's think twice	30
The balance of nature	31
To kill or not to kill?	31
Index	32

CREEPERS AND CRAWLERS

THE creatures in this book belong to a group of animals called arthropods. They are small and boneless. Their bodies are divided into segments, or sections, and are covered by a tough shell.

Insects are six-legged arthropods. Ticks, spiders, and other arachnids are eight-legged arthropods. Centipedes and millipedes are arthropods with hundreds of legs!

 Arthropod means "jointed legs."

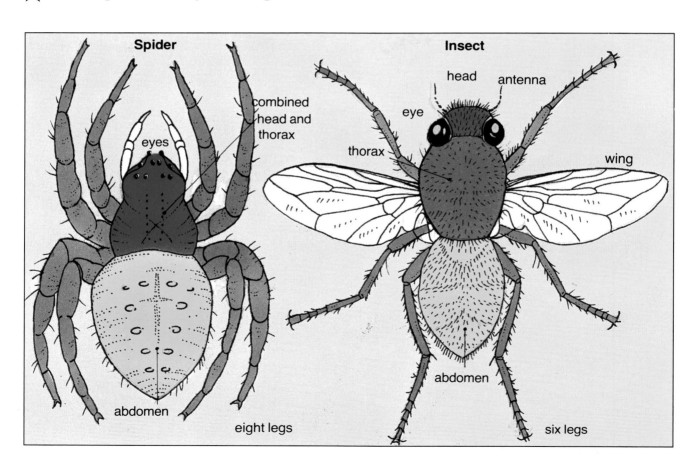

BITERS, CHEWERS, AND SUCKERS

ARTHROPODS eat anything. Some eat flesh or suck blood. Others eat nectar, sap, or plants. Many eat their own kind. Their mouthparts vary, depending on the type of food they eat. Arthropods that bite have powerful jaws called mandibles. Those that suck have a long, thin tube called a proboscis.

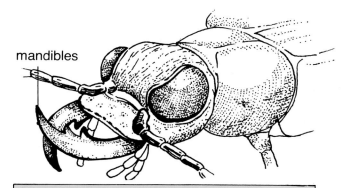

SMALL BUT DEADLY

A TIGER or a shark can attack and kill humans. However, insects and spiders are tiny. They cannot be dangerous, can they?

The fact is that creatures such as insects have been on Earth for about 350 million years. Many have developed weapons in order to survive. They may sting or bite, injecting poison into the blood stream.

Some are dangerous in other ways. They may spread disease, or they may destroy crops, which then causes starvation.

Praying mantis

Who rules the world?

We often think that humans rule the world. However, there are more than a million different types, or species, of insects. Just 2½ acres of farmland may contain nearly 200,000 million insects! As a group of animals, insects have conquered the land, the air, and the water.

YOUNG AND DANGEROUS

INSECTS change their shape at different stages of their lives. They start their lives as eggs. When they hatch, they are called larvae. The larvae of different insects are known as caterpillars, grubs, or maggots. As the larvae become bigger, they outgrow their hard outer shell and must change it.

Some insects then become adults. Others must go through one more stage. Then they are known as pupae.

Young insects need huge amounts of food to keep growing. The young, or nymphs, of the dragonfly live in ponds. They are fierce hunters. Others, such as the caterpillars of many butterflies and moths, attack crops. Some eat so much that they can ruin farmers and even cause famine.

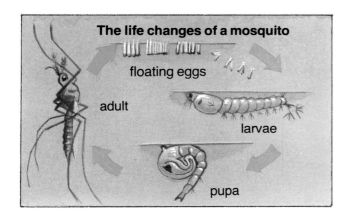

The life changes of a mosquito

SPOILERS OF FOOD

MANY larvae get into stored food. Because they make it unfit for eating, they can cause hunger. Caterpillars of the mealworm and the Mediterranean flour moth feed on stored flour and grain. Caterpillars of the tobacco moth eat cocoa beans. Many flies lay their eggs in meat.

Years ago, sailors on long voyages had to eat food that was spoiled. Today, packaging, refrigeration, and freezing helps protect our food.

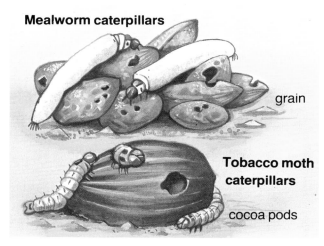

POISON CATERPILLARS

MANY eggs, caterpillars, and pupae of butterflies and moths are eaten by birds or by other insects. Some use poison to protect themselves.

The puss moth caterpillar spits formic acid. The caterpillar of a Venezuelan emperor moth causes an illness that prevents blood from clotting.

 Watch out for hairy caterpillars. In some, a poison is released when the hairs are touched or broken. These can cause painful skin rashes.

Puss moth caterpillar

DEADLY STINGERS

BEES, wasps, and hornets are the most dangerous of all stinging insects. They may kill more people than snakes do. About 4,000 people die from bites each year.

Many people are allergic to the poison, or venom. The effects of the sting are worse for them than for others. People stung in the throat can suffocate when the swelling begins.

Sometimes a swarm of bees attacks one person. One victim was stung more than 2,200 times by a single swarm – and lived!

Common wasp

WHY DO WASPS AND BEES STING?

NORMALLY, wasps and bees will only sting humans in self-defense. Some species of wasp sting in order to paralyze or kill their prey. Others sting caterpillars so that their own larvae will have fresh meat to eat as they grow.

 Only female bees can sting. The sting has developed from the egg-laying organ.

 A bee uses 22 muscles when it stings!

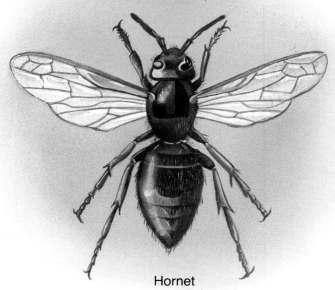
Hornet

How does a bee sting work?

The sting is made up of three barbed daggers grouped around a canal. When the bee senses danger, it activates the sting. The sting is pushed out of its cover, or sheath, and jabbed into its victim. Venom flows down the canal and into the victim.

A bee sting

WARNING COLORS

MANY bees and wasps are brightly colored, with black, yellow, or red stripes. These colors warn of danger. Some harmless insects have copied these colors. In this way, they are left alone, too!

KILLER BEES

THE most aggressive stinger of all may be the African bee. Years ago, they were imported into South America. Some escaped and bred with local bees. The offsprings are just as fierce. They are spreading northward.

African bee

SOLDIERS ON THE MARCH

ANTS are relatives of the bees and wasps. They range in length from 1/10 to 1 inch. There are about 10,000 different species. The females of many species can bite and sting. Some wood ants can squirt formic acid at an attacker.

Wood ant

ARMY LIFE

ANTS live in huge nests, ruled by a queen. Each ant has its own job to do. Female workers gather food. Some species have large female soldiers that defend the nest. They are very aggressive. If an ant from another nest wanders in, it is killed at once. In an ant's nest, the ants are always on the alert.

Killer venom

The most dangerous ant in the world is the black bulldog ant of Australia. It stings and bites. It can kill a human in 15 minutes.

Black bulldog ant

COLUMNS OF DEATH

IN South America you may see hundreds of thousands of army ants marching in long lines. In Africa you may see driver ants on the march. They eat anything in their path, alive or dead. Humans can usually escape easily enough. However, babies and injured adults may be in danger.

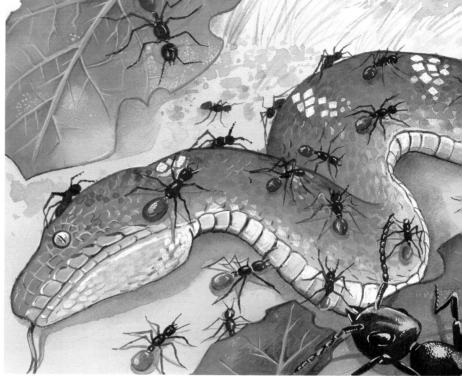

Army ants

Termite mound

THE DEMOLITION SQUAD

TERMITES look like ants, but they are not related. They lead similar lives, and are ruled by a king and a queen. Large termites called soldiers defend the nest. The nest may be underground, or it may be a huge mound or tower of earth.

Termites do not eat flesh. They eat wood. They can infest whole buildings, which may eventually crumble and collapse.

TIRELESS JAWS

BEETLES form the largest group of insects. There are more than 250,000 species. They are greedy eaters that have powerful jaws. Some species eat plants. Others eat flesh. Many are fierce hunters. They rarely attack humans, but they are still a threat.

THE POTATO-EATERS

MANY beetles attack forests and crops. The Colorado potato beetle is less than a half inch long. The female lays her eggs on the leaves of the potato plant. These hatch into grubs. Both grubs and adults attack the plants. They can strip entire fields.

Colorado beetle

The Colorado beetle originally came from the western parts of the United States. There it ate a plant called buffalo burr. As farmers settled the land and planted fields of potatoes, the beetle began to eat these plants. It spread eastward in the last century, destroying potato crops. By the 1920s it had reached Europe. It had been carried to France in a ship's cargo.

POISON BEETLES

THE bodies of some beetles contain a poison called cantharidin. This can burn and blister human skin, and hurt very much.

Some blister beetles are brightly colored. They are between 1/8 and 1 inch long. Most blister beetles give off a bad-smelling liquid when they are attacked. It comes from between the joints of their legs. The liquid causes redness and blisters when it touches the skin.

Blister beetle

KITCHEN PESTS

COCKROACHES look like beetles, but they are not related. There are about 3,500 species. Some species grow more than 3 inches long. Most of them live in warm countries. They feed on dead animals, rotting food, and scraps. Some kinds have moved into buildings. They thrive in the warmth of kitchens. There, they spoil stored food and can spread food poisoning and disease.

Cockroach

DESTROYERS

THE most destructive insect is the desert locust of Africa, the Middle East, and India. Locusts are large grasshoppers that live in tropical countries. They damage so many crops that they cause widespread famine.

CROP STRIPPERS

GRASSHOPPERS usually live on their own. However, every so often locusts gather in large numbers called swarms. When they find a good feeding place, they lay their eggs. These hatch into young hoppers.

Desert locust

LIFE IN THE SWARM

WHEN hoppers grow into adults, they take to the skies. They live as a swarm over several generations. During this phase they look different than locusts that live on their own. A large swarm may have up to 250,000 million insects. It can cover hundreds of square miles in search of food.

Farmers fear the coming of the swarm. Although they try to keep away the insects, little can be done. A locust swarm can eat 3,000 tons of leaves in a day. When the locusts run out of food, they die.

The semi-desert regions where the locusts live are hard to farm. The coming of the locusts can mean starvation for poor villagers.

PEST CONTROL

FOR years locusts have been sprayed and destroyed at their breeding grounds. However, swarms still occur and it is hard to predict where they will go.

DIRT BREEDS DANGER

THE most dangerous insect of all is the housefly. It lays its eggs in waste matter and garbage. The adult housefly does not bite, but it does drool on its food and spread disease.

Its long mouthparts form a bent proboscis, with wide tips for licking and sucking. As it feeds, saliva carries digested food back down the proboscis. This helps the fly digest its next mouthful. This method of feeding spreads germs rapidly. To keep flies away, food should be covered.

★ **The housefly may be responsible for passing on 30 different illnesses or infections. These include several deadly diseases.**

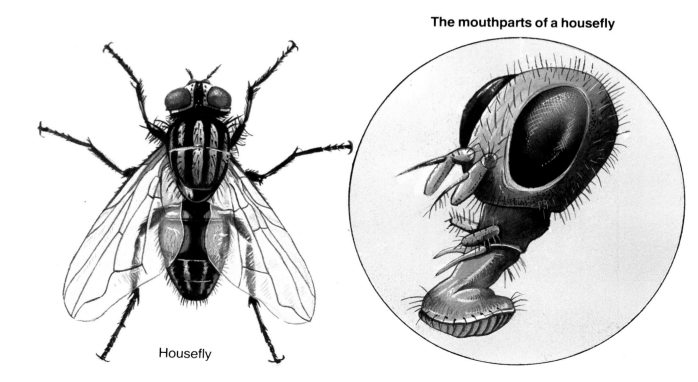

Housefly

The mouthparts of a housefly

MAGGOT-BREEDERS

SPECIES of blowfly include the bluebottle and the greenbottle. The bluebottle buzzes around kitchens. It lays its eggs in meat or fish. The grubs, or maggots, hatch within a day.

The greenbottle lays its eggs on dead animals. Sometimes it lays them on living sheep. When its maggots hatch, they burrow into the flesh. This causes festering sores.

Maggots

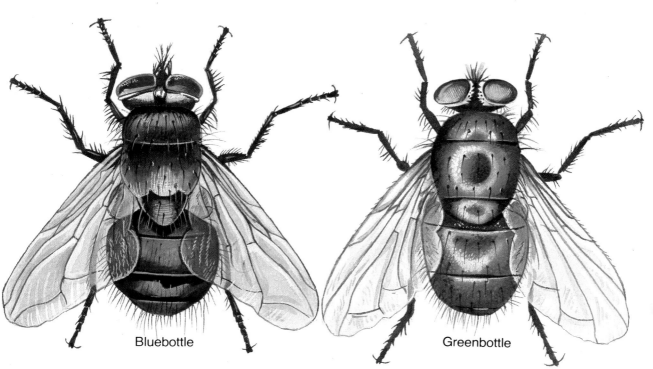

Bluebottle Greenbottle

BLOODSUCKERS

MOSQUITOES are two-winged flies. They have slender bodies and long legs. Most are less than a half inch long. They feed upon nectar.

There are about 3,000 mosquito species. Many of the females must suck blood from a bird or a mammal before they can lay eggs. Their mouthparts are designed to do this. They form long needles called stylets. These come out of the proboscis and are injected into the skin. A chemical prevents the blood from clotting. It is this that causes the pain and swelling in a mosquito bite. Some people are allergic to the chemical.

Mosquito larvae hatch in pools of water. Chemicals are sprayed into the pools in order to kill the insects.

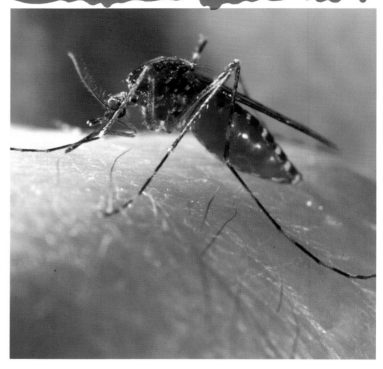

Disease carrier

The bite of a mosquito is not really dangerous. What is dangerous is the diseases some tropical mosquito species pass along.

These diseases include yellow fever and malaria. Perhaps a million people die from diseases carried by mosquitos each year. Millions more become very ill.

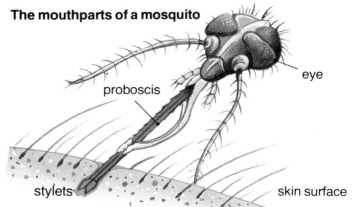

The mouthparts of a mosquito

THE SCOURGE OF AFRICA

THERE are 20 species of tsetse fly in Africa They resemble blowflies. However, they have a needlelike proboscis for sucking blood. Like the mosquito, the tsetse injects its victims with a chemical to make the blood keep flowing. Both males and females are blood-suckers.

Some tsetse flies pass on a tiny parasite, or a creature that lives off another one. This organism, called trypanosome, lives in the bloodstream of other creatures. It causes an animal disease called nagana. This can affect whole herds of cattle. It also causes a deadly human disease called sleeping sickness. Tsetses are still very hard to control.

BUGS AND LICE

ALL kinds of tiny parasites live off the human body. Human head lice grip on to hair, and suck blood from the scalp. They lay eggs called nits in hair. Human body lice live in clothes and feed on the body.

Anyone can pick up lice. Fortunately, they do little lasting harm if treated. Hair must be washed with a special shampoo. Clothes must be cleaned and changed regularly.

Head louse

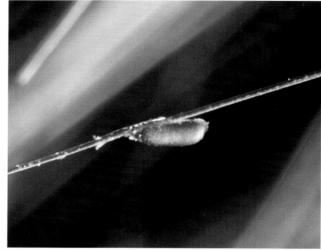

Nit

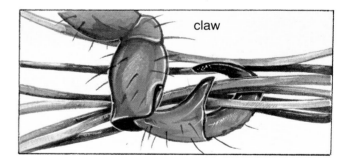

claw

Lice and disease

Two or three hundred years ago, homes were crowded and people did not wash themselves or their clothes often. Lice were a serious problem. They carried deadly diseases like typhus.

STRANGE BEDFELLOWS

THE bedbug infests houses. It lays its eggs in cracks and behind wallpaper. When it hatches, it often moves to a bed. There it can suck blood for its food.

There are two species that attack humans. One kind lives in Europe and North America. The other lives in Asia and Africa. The bedbug has a kind of beak called a rostrum that contains needles, or stylets. These suck blood and inject a chemical that irritates skin.

Bedbug

THE PLAGUE FLEAS

THERE are about 1,500 species of flea. These tiny parasites live on the bodies of mammals and birds. Many fleas only live on one kind of animal. Others live on two or more species. The human flea, for example, also lives on pigs, foxes, and badgers. It has powerful legs and can jump 12 inches from one body to another. It is a bloodsucker. Its mouthparts form a needle.

The oriental rat flea usually lives on rats. It feeds on their blood. Sometimes it picks up a terrible germ from the rats. When the rat dies of the germ, the flea may jump on to humans and start biting them. This gives the person a disease called bubonic plague. The victim dies unless treated with drugs.

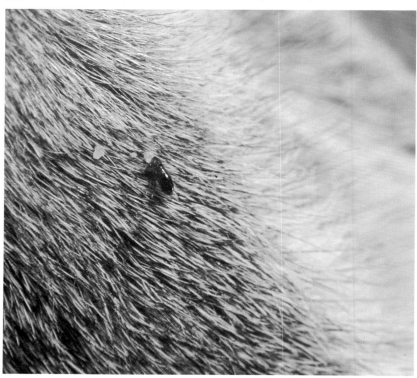

Rabbit flea

★ **About ten million people died of the plague in India between 1896 and 1917.**

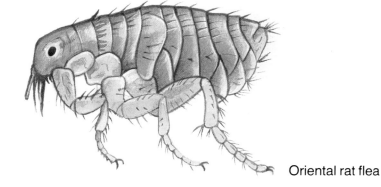

Oriental rat flea

A DEADLY DISEASE

IN the olden days, cities were full of rats, and rats were full of fleas. There were no sewers or garbage collections, and no safe rat poisons. When the plague broke out, the effects were terrible.

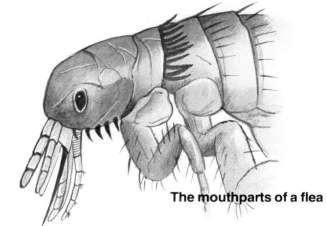

The mouthparts of a flea

★ **In 1347 the Black Death arrived in Europe from Asia. It killed between one-third and half of the population. It returned again and again.**

★ **London was stricken again between 1563 and 1665. Hundreds of thousands of people were killed. The corpses were buried in mass graves.**

Because of the bubonic plague, the tiny oriental rat flea is the most feared of all insects. Fortunately the disease is now very rare, although it sometimes still occurs in the Far East. Cities are now cleaner and rats are controlled. Doctors understand how the disease works, and there are medicines to help the victims.

POISON SPIDERS

MANY people fear spiders. Although most of the 32,000 known species of spider are venomous, few are a threat to humans. The spiders that are dangerous usually bite only in self-defense. Then, they can kill.

Eight legs and poisonous fangs

Spiders are not insects. They are arachnids with eight legs. They are very numerous and eat large amounts of other insects. They may run after their prey and kill it. They may also trap it in a web of silken thread. All spiders seize their prey and then bite it with their fangs. These are usually connected to a gland that produces venom. Some female spiders eat their own mate; and some baby spiders eat their own mother.

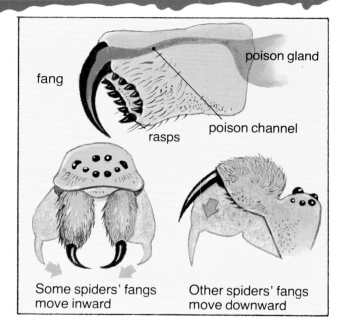

Some spiders' fangs move inward

Other spiders' fangs move downward

Red-kneeded bird-eating spider

HAIRY HEAVYWEIGHTS

THE bird-eating spiders are the biggest of all arachnids. They live in South and Central America. They eat insects, but they are large enough to eat a hummingbird or a small mammal, as well. Their legs may be 10 inches apart, and their fangs can be 1 inch long.

Humans are not really in danger from these giants. However, their bite is painful. Bird-eating spiders are very hairy. Touching the hairs may irritate the skin.

THE DEADLIEST OF ALL

THE female black widow spider of North America and the Caribbean may be the most dangerous spider of all. Even its eggs are poisonous. Its venom is much deadlier than the venom of a rattlesnake.

The black widow is not aggressive. However, if it is touched, it may bite. The venom causes sickness, paralysis, madness, and sometimes death. Relatives of the black widow are found on all continents.

★ **The Phoneutria spiders of Brazil have larger venom glands than any other spider. They hide in shoes and clothes. Attacks are common. Young children are in the greatest danger unless medical aid is given quickly.**

★ **The term "tarantula" is often used to describe any large and hairy spider. It should only refer to a wolf-spider from southern Europe. Its bite was once thought to make its victims shake. A wild dance called the tarantella was named after it.**

Female black widow spider

A STING IN THE TAIL

THERE are about 1,200 species of scorpion. They feed mostly on insects and spiders. Like spiders, scorpions are arachnids and have eight legs. They vary in size from less than a quarter inch to 8 inches.

Scorpions are found in the world's warmer regions. They have two front pincers, which grab prey. Their jaws tear the flesh apart. The scorpion's secret weapon is in its tail. On the tip is a sting that injects venom.

Scorpions usually sting in self-defense.

THE POISONERS

THE most venomous scorpion is *Leiurus quinquestriatus*. It lives in North Africa. Fortunately, it only injects a tiny amount of venom. Otherwise, many more people would become its victim. As it is, babies and young children run a high risk of being stung to death.

Androctonus australis is another North African arachnid. Sometimes it is called the fat-tailed scorpion. It is probably the most dangerous of all. Records show that it has killed hundreds of people. Other killer scorpions are found in southern Africa, the Caribbean, Central and South America, and the southwestern United States.

Leiurus quinquestriatus

Fat-tailed scorpion

The blue death
Most scorpion stings are not fatal. Quick treatment can usually save a victim. People who have been stung become sick, and break into a sweat. They cannot speak or breathe properly. Their body begins to twitch, and their arms and legs turn blue. They may die within a few hours.

INSECT FRIENDS

ALTHOUGH many insects are deadly, we should be able to live peacefully with most of them. We depend on insects for our survival.

Insects spread pollen from one plant to another, helping our orchards to produce fruit. Bees give us honey. Many insects, like the ladybug, kill other kinds of insects that are harmful to us. Many insects are a source of food for birds, mammals, and other creatures. A few are eaten by humans.

LET'S THINK TWICE

WE should think carefully before we call insects killers. We often spray chemicals on crops from planes. The wind may blow those chemicals in all directions. Friendly insects may be killed along with pests. The amount of poison in an insect's sting is tiny by comparison.

A hundred years ago, people killed butterflies and other insects for their collections. We must still collect insects for research, but not merely for a hobby.

THE BALANCE OF NATURE

M ANY species of insects are becoming rare, or have died out. This may be because the climate is changing. It may also be because of the chemicals farmers sprayed on crops, or because we changed the areas where insects live.

Living things depend on each other. If we spray a crop to kill insects, we may also poison the birds and animals that feed on them.

TO KILL OR NOT TO KILL?

W E must control the spread of disease and destroy the breeding grounds of locusts and mosquitoes. We must make sure our crops grow well in order to prevent famine. However, if we continue to fill the land with too many chemicals, we may poison the whole world.

Arthropods are survivors. Maybe we can learn from them.

INDEX

Africa 13, 16, 21, 23, 29
African wild bee 11
Androctonus australis 29
ant 12, 13
arachnid 6
army ants 13
arthropod 6, 7
Asia 23, 25
Australia 12

bedbug 23
bee sting 10, 11, 26
bees 10, 11, 12, 30
beetles 14, 15
bird-eating spider 26
birds 9, 26, 30, 31
black bulldog ant 12
Black Death 25
black widow spider 27
blister beetle 15
blood 7, 20, 21, 22, 23
bluebottle 19
Brazil 27
breeding grounds 17, 31
bubonic plague 24, 25
buffalo burr 14
buildings 13, 15, 23
butterflies 8, 9, 30

cantharidin 15
Caribbean 27, 29
caterpillars 8, 9, 10
  poisonous 9
cattle 21
centipedes 6
chemicals 20, 23, 30, 31
cockroaches 15
Colorado potato beetle 14
crops 7, 8, 14, 16, 30, 31

danger colors 11
death 16, 29, 30
desert locust 16
disease 7, 15, 18, 20, 24, 25
doctors 15, 25, 27, 29
dragonfly 8
driver ants 13

eggs 8, 9, 14, 16, 18, 19, 20, 23
Europe 14, 23, 25, 27

famine 8, 16
fangs 26
Far East 25
farmers 8, 14, 17
fat-tailed scorpion 29
flea 24, 25
flies 9, 18, 19, 20, 21
food poisoning 15
formic acid 9, 12
France 14

garbage 18, 25
germs 18, 24
grain 9
greenbottle 19
grubs 8, 14, 19

hornets 10
housefly 18
human body lice 22
human head lice 22

illness 9, 18
India 16, 25
insects
  biting 12, 20, 21, 22, 23, 24, 25, 26
  disease carrying 15, 18, 19, 20, 21
  life cycle 8
  poisonous 9, 15, 28, 29, 30
  stinging 10, 11, 12, 28, 29, 30
  swarming 10, 17

jaws 7, 14, 28

kitchens 15, 19

larvae 8, 9, 10, 18, 20
Leiurus quinquestriatus 29
lice 22
locusts 16, 31

maggots 8, 19
malaria 20
mandibles 7
mealworm 9
meat 9, 10, 14
medicine 24, 25
Mediterranean flour moth 9
Middle East 16
millipedes 6
mosquitoes 20, 31
moths 8, 9

nagana 21
nectar 7, 20
nits 22

North America 23, 27
nymphs 8

oriental rat flea 24, 25

parasite 21, 22, 23, 24
Phoneutria spider 27
plague 24, 25
plants 7, 14
potato 14
proboscis 7, 18, 20, 21
pupae 8, 9
puss moth caterpillar 9

rash 9
rats 24, 25
rostrum 23

scorpions 28, 29
self-defense 10, 28
sewers 25
skin 9, 15, 23, 26
sleeping sickness 21
sores 19
South America 11, 13, 26, 29
spiders 6, 7, 26, 27, 28
stylets 20, 23
suffocation 10

tarantella 27
tarantula 27
termites 13
ticks 6
tobacco moth 9
trypanosome 21
tsetse fly 21
typhus 21

United States 14, 29

venom 10, 11, 26, 27, 28, 29, 30
Venuzuelan emperor moth 9

wasps 10, 11, 12
wolf spider 27
wood ants 12

yellow fever 20